YOU WILL PASS GEOMETRY

You Will Pass Geometry

Poetry Affirmations for Math Students

Walter the Educator™

SKB

Silent King Books a WhichHead Imprint

Copyright © 2023 by Walter the Educator™

All rights reserved. No part of this book may be reproduced in any manner whatsoever without written permission except in the case of brief quotations embodied in critical articles and reviews.

First Printing, 2023

Disclaimer
This book is a literary work; poems are not about specific persons, locations, situations, and/or circumstances unless mentioned in a historical context. This book is for entertainment and informational purposes only. The author and publisher offer this information without warranties expressed or implied. No matter the grounds, neither the author nor the publisher will be accountable for any losses, injuries, or other damages caused by the reader's use of this book. The use of this book acknowledges an understanding and acceptance of this disclaimer.

dedicated to all the math lovers across the world

CONTENTS

Dedication v

Why I Created This Book? 1

One - Embrace Geometry 2

Two - Will Open The Door 4

Three - Ignite Your Soul 6

Four - Conquering Geometry 8

Five - Unlock The Doors 10

Six - Passion And Zeal 12

Seven - Strong And Enough 14

Eight - Setting Your Mind Free 16

Nine - Diligence And Might 18

Ten - Your Guiding Light 20

Eleven - Endless Possibility 22

Twelve - Believe In Yourself 24

Thirteen - Future Shining Bright 26

Fourteen - Wonders Of The World 28

Fifteen - Guide Your Quest 30

Sixteen - Geometry's Secrets 32

Seventeen - Geometry's Wisdom 34

Eighteen - Knowledge Intertwine 36

Nineteen - You'll Find The Key 38

Twenty - Power To Shape 40

Twenty-One - Triangles, Circles, And Polygons Unite 42

Twenty-Two - Illuminate Your Way 44

Twenty-Three - Pure Delight 46

Twenty-Four - Forever Explore 48

Twenty-Five - Dear Student 50

Twenty-Six - Seize The Day 52

Twenty-Seven - Geometry, Inspire 54

Twenty-Eight - Soul Collide 56

Twenty-Nine - Lessons Are Profound 58

Thirty - Labyrinth Of Shapes 60

Thirty-One - Study With Passion 62

Thirty-Two - Dedication 64

Thirty-Three - Brighten Your Day 66

Thirty-Four - Never Pass You By 68

Thirty-Five - Geometry's Rewards 70

About The Author 72

WHY I CREATED THIS BOOK?

Creating a poetry book to motivate a student to pass the subject of Geometry was an effective way to engage their interest and make the learning process more enjoyable. Poetry has a unique ability to convey complex concepts in a creative and memorable way, making it a valuable tool for education. By using poetic language and imagery, this book can demystify complicated geometric principles, providing a fresh perspective and encouraging the student to explore the subject with enthusiasm. Additionally, this book can instill a sense of confidence and accomplishment, boosting the student's motivation to succeed in Geometry.

ONE

EMBRACE GEOMETRY

In the realm of angles and lines,
Where shapes and figures intertwine,
Lies the subject of Geometry,
A puzzle waiting to be set free.
    Oh, student, let me be your guide,
Through this maze of the geometric stride.
Discover the beauty that lies within,
As you delve into this world of kin.
    For Geometry, my dear friend,
Is more than just a means to an end.
It teaches logic, reasoning, and grace,
A foundation for life's intricate space.
    With Euclid as your guiding light,
You'll conquer triangles, circles, and right.

Unlock the secrets of Pythagoras' theorem,
And watch your understanding gleam.
    From polygons to theorems profound,
Geometry's wonders will astound.
Let angles and lines become your friends,
As you journey towards knowledge that transcends.
    So, fear not the challenge that Geometry brings,
Embrace it with passion, let your heart sing.
For within its depths, you'll find the key,
To a world of possibilities, both vast and free.
    Let your curiosity lead the way,
As you strive to conquer each display.
And as you master this subject's art,
You'll find the joy that comes from a fresh start.
    So, dear student, let me implore,
Embrace Geometry and strive for more.
For in the realm of angles and lines,
Lies the power to shape your own designs.

TWO

WILL OPEN THE DOOR

In the realm of shapes and lines, behold,
The world of Geometry, a tale untold.
Where logic and reason intertwine,
A subject that transcends the confines of time.
    Oh, student, fear not this geometric land,
For within its grasp, knowledge grand.
With compass and ruler, you shall explore,
The secrets of angles and polygons galore.
    Let it be known, dear student, the power it holds,
Geometry unlocks wonders, untold stories unfold.
For in its depths lies a grace divine,
Where symmetry and beauty forever entwine.
    With Pythagoras as your guiding light,
Discover the harmony, the balance so bright.

Through axioms and theorems, you shall tread,
Unraveling the mysteries, no longer misled.
  Embrace the challenge, let your mind soar,
Geometry's lessons will open the door.
To a world of possibilities, vast and wide,
Where you shall shape your own designs with pride.
  So strive, dear student, with all your might,
For Geometry's treasures are within your sight.
Unlock the secrets, let your passion ignite,
And triumph over every geometric fight.
  For in the realm of shapes and lines, you'll see,
Geometry's magic will set your spirit free.
With logic and reasoning as your guide,
Success in this subject shall be your pride.

THREE

IGNITE YOUR SOUL

In the realm of shapes and lines,
Where wonders of Geometry unwind,
A student stands, determined and bold,
Seeking knowledge, a treasure to behold.

Geometry, a symphony of angles and curves,
Unveiling the mysteries that the universe preserves,
From circles to triangles, polygons too,
Each figure holds a story, waiting to be pursued.

Fear not the theorems and postulates,
For they are the keys to unlock the gates,
To a world where patterns and logic combine,
Where imagination soars and dreams align.

Embrace the compass, the ruler, and the square,
They are your allies, guiding you with care,
Measure, calculate, and analyze,
And watch as your understanding multiplies.

For in Geometry lies the power to create,
To design, to innovate, to elevate,
From architecture's grandeur to nature's grace,
Geometry weaves its magic in every space.

So let the beauty of shapes ignite your soul,
Let the allure of angles make you whole,
With passion and perseverance, you will succeed,
And Geometry's wonders you will exceed.

For within its realm, a universe unfurled,
Geometry, the gateway to a boundless world,
Unlock its secrets, let your dreams take flight,
And conquer the realm of shapes with all your might.

FOUR

CONQUERING GEOMETRY

In the realm where lines and angles intertwine,
Geometry's magic begins to shine.
A subject of shapes, both simple and complex,
It holds the key to unlocking life's perplex.

For every curve and corner, a mystery lies,
A puzzle to solve, a truth to realize.
Oh, student, fear not this mathematical art,
Embrace the challenge, let it ignite your heart.

From Pythagoras to Euclid's ancient grace,
Geometry weaves a tapestry in space.
It's the language of nature, the blueprint divine,
The symphony of patterns that make life align.

Triangles, circles, and polygons galore,
Geometry's playground, where wonders explore.

Discover the secrets of symmetry's dance,
And embrace the beauty of every circumstance.
    For in the realm of Geometry's embrace,
Lies the power to soar, to reach infinite space.
So, gather your compass and straight-edge with pride,
For in conquering Geometry, you'll surely stride.
    Let passion guide you, let knowledge unfold,
Geometry's treasures are waiting to be told.
Unlock the doors of possibility's maze,
And witness the brilliance that Geometry displays.
    So, fear not, dear student, take your rightful place,
Embrace Geometry's beauty and win the race.
With dedication and perseverance, you'll find,
Geometry's gift will forever be kind.

FIVE

UNLOCK THE DOORS

In the realm of shapes and lines,
Where logic and beauty intertwine,
Lies the realm of Geometry,
A subject of grace and mystery.

Oh student, hear my heartfelt plea,
Unlock the doors to Geometry.
For in its depths, you'll surely find,
A path to expand your brilliant mind.

Geometry teaches us to see,
The world around with clarity.
From circles round to angles sharp,
It helps us navigate life's arc.

Through triangles and polygons,
We learn to reason, build upon.

With compass and ruler in our hand,
We shape our world, we take a stand.
    So fear not Pythagoras' theorem,
Or theorems of Euclidean dream.
For in the realm of Geometry,
Lies the power to set your mind free.
    Embrace the challenge, rise above,
Discover the wonders you'll come to love.
For in the realm of shapes and lines,
You'll find the strength to pass with shine.
    So let your passion ignite the flame,
And conquer Geometry's noble aim.
For in its grasp, you'll surely see,
The beauty of logic and harmony.

SIX

PASSION AND ZEAL

In the realm of shapes and lines,
Where logic and beauty intertwine,
Lies the subject of Geometry,
A gateway to endless possibility.

Let us delve into this wondrous art,
That teaches us to think and impart
The power of reason, the grace of design,
To unravel the mysteries hidden in time.

Geometry, the language of space,
Unveils the secrets with elegant grace.
Angles and curves, triangles and squares,
Building blocks of knowledge, beyond compare.

For in the realm of Geometry's embrace,
We find the puzzles that challenge and chase,
The mind to expand, to grow and explore,
Unlocking the wonders never seen before.

So, dear student, fear not the unknown,
Embrace the challenge, let your mind be blown.
Geometry beckons, with its mystical charm,
To shape your thoughts and keep you warm.

With every theorem, every proof,
Your mind will soar, like a graceful roof.
And when you pass this sacred test,
A world of possibilities will manifest.

So, study hard, with passion and zeal,
For Geometry's secrets it will reveal.
And as you conquer each geometric feat,
Success will be yours, oh student sweet.

SEVEN

STRONG AND ENOUGH

In the realm where angles dance,
And shapes weave a graceful trance,
Lies a subject oft misunderstood,
Geometry, the key to a world so good.

Through the lines that intersect and meet,
Lies the power to unlock logic so sweet,
With triangles, circles, and polygons too,
Geometry teaches us what's real and true.

In every corner, a theorem waits,
To challenge minds, to test our fates,
But fear not, dear student, for in this quest,
Lies the strength to conquer any test.

For Geometry, you see, is more than shapes,
It teaches us to think, to reason, to create,

With every problem, a solution unfurled,
Geometry guides us through a logical world.
　So study hard, with passion and grace,
Embrace the challenge, seek knowledge's embrace,
For Geometry's lessons, though they may seem tough,
Will shape your mind, make you strong and enough.
　For in this subject lies a secret sublime,
The beauty of patterns, the wonders of time,
So let your curiosity soar, let your spirit be free,
Geometry's mysteries, the world awaits to see.

EIGHT

SETTING YOUR MIND FREE

In the realm of shapes and lines, behold,
Geometry's beauty, a tale untold.
A subject that may seem complex and vast,
Yet holds within it a power unsurpassed.

    Geometry, the language of form and space,
Teaches logic, reasoning, and grace.
With angles and proofs, it challenges the mind,
Unlocking the secrets we're destined to find.

    Through triangles, circles, and polygons too,
Geometry reveals a world that's true.
In every curve and every straight,
A symphony of patterns, a masterpiece awaits.

    Embrace the challenge, dear student, with zeal,
For Geometry's wonders are truly surreal.

Let its puzzles and theorems ignite your fire,
And fuel your passion to reach higher.

    For within the realm of Geometry's embrace,
You'll find the magic, the beauty, the grace.
So study well, persevere with might,
And conquer the realm where shapes take flight.

    Geometry's realm, a gateway it shall be,
To a universe of infinite possibility.
Pass this subject, and you'll surely see,
The wonders of Geometry, setting your mind free.

NINE

DILIGENCE AND MIGHT

In the realm of shapes and lines, where angles intertwine,
Lies the treasure trove of knowledge, where brilliance does shine.
Geometry, the gateway to logic and reason,
A subject that unveils the secrets of nature's season.

    Oh student, listen closely, for I shall impart,
The beauty of Geometry, it stirs the heart.
With compass and ruler, we navigate the plane,
Unlocking the truths that our minds can attain.

    Geometry teaches us grace, elegance, and poise,
As we dance with triangles, circles, and noise.
The symmetry and patterns, like music to our eyes,
We learn to see the world through geometric ties.

Through polygons and polyhedra, we explore,
The wonders of dimensions, forever to adore.
From Pythagorean theorem to the golden ratio,
Geometry holds the key to wisdom's bravado.

So study, dear student, with diligence and might,
For Geometry is a beacon, a guiding light.
In its embrace, you'll find the power to excel,
And conquer the subject that you once thought fell.

Embrace the challenges, let curiosity soar,
For Geometry's magic, you can't ignore.
With every theorem, every proof you'll see,
The wonders of Geometry, and the wonders in thee.

TEN

YOUR GUIDING LIGHT

In the realm of shapes and lines, where angles dance and curves entwine,
Lies the key to unlock the gate, where wisdom and knowledge intertwine.
Geometry, a language of grace, speaks to us with elegance and finesse,
Guiding us through the labyrinth of space, where beauty and logic coalesce.
Oh, student brave, with mind aflame, fear not the challenge that Geometry brings,
For in its depths, a treasure awaits, like a diamond that forever sings.
With compass and ruler as your guide, you'll navigate the paths unknown,
Discovering truths that will forever abide, in the chambers of your heart and your bone.

Geometry, the art of pure thought, unveils the mysteries of the universe's design,

It teaches us to reason, to ponder, to seek answers that lie beyond the line.

Embrace its puzzles, its theorems, its proofs, for within them lies a world untold,

Where imagination soars and logic aloofs, where dreams and reality unfold.

So, dear student, with courage renewed, let your passion for Geometry ignite,

For within its web lies a power pursued, a brilliance that shines ever bright.

Unlock the gates of your own mind's door, let Geometry be your guiding light,

And in your journey, forevermore, may you find wisdom's eternal flight.

ELEVEN

ENDLESS POSSIBILITY

In the realm of angles and lines,
Where shapes and patterns intertwine,
Lies the secret, so divine,
Of Geometry, a subject so fine.
    Oh, student, let me tell you true,
Geometry holds a gift for you,
Beyond theorems and equations, see,
It teaches logic, reasoning, and grace, with glee.
    For in the dance of lines, you'll find,
A world where logic intertwines,
Reasoning, sharp as a knife,
Will guide you through the maze of life.
    Geometry, the architect's delight,
Builds bridges to the stars at night,

With shapes and forms, it seeks to find,
The beauty that lies in every line.

    So, dear student, don't despair,
Embrace the challenge, if you dare,
For Geometry holds the key,
To unlock the wonders that lie within thee.

    Pass this subject, with heart and zest,
And you'll discover, among the rest,
That Geometry's embrace will be,
A gateway to a world of endless possibility.

TWELVE

BELIEVE IN YOURSELF

In the realm of shapes and lines,
Where logic's dance forever shines,
Lies the world of Geometry,
A subject that holds the key.

Oh, student, do not despair,
For in this realm, you'll find a stair,
That leads you to a higher plane,
Where knowledge and grace do intertwine.

Geometry, a language pure,
Teaches you to think and endure,
To reason with clarity and precision,
And unlock the secrets of this mission.

From circles to triangles, polygons too,
Each shape reveals a truth to you,

Angles, arcs, and parallel lines,
Formulating patterns, oh so divine.
    Embrace the challenge, seek the thrill,
Geometry's wonders will fulfill,
For within its depths lies a hidden treasure,
A world of shapes, beyond all measure.
    So, study hard, with passion and zest,
And let Geometry put you to the test,
For by mastering this subject's grace,
You'll find a path to conquer any space.
    Believe in yourself, the journey is near,
Geometry's wisdom will soon appear,
And as you pass with flying colors,
A world of possibilities shall uncover.

THIRTEEN

FUTURE SHINING BRIGHT

In the realm of shapes and lines,
Where logic and beauty intertwine,
Lies the subject of Geometry,
A gateway to a world of clarity.

Oh, dear student, do not despair,
Embrace this challenge, if you dare,
For Geometry holds treasures untold,
Waiting to be discovered and unfold.

Within its angles and measurements,
Lies the power of logical assessments,
With compass and ruler in hand,
You'll unlock the secrets of this land.

Geometry teaches us to reason,
To think critically, beyond mere season,

It sharpens our minds, opens doors,
Guiding us on paths we've never explored.

Through symmetry and congruence,
We find grace in every occurrence,
From pyramids to the stars above,
Geometry reveals the wonders we love.

So, study hard and never give in,
For Geometry is a journey to begin,
With perseverance and determination,
You'll conquer this subject, no hesitation.

And when you pass with flying colors,
You'll realize it's not just about numbers,
Geometry imparts wisdom and might,
A foundation for a future shining bright.

FOURTEEN

WONDERS OF THE WORLD

In the realm of shapes, where lines converge,
Lies the key to unlock the knowledge you deserve.
Geometry, a subject of logic and grace,
Teaches you to navigate life's intricate space.

With triangles, circles, and angles galore,
You'll learn to reason and explore.
Through the language of shapes, you'll find,
A world of connections that intertwine.

Geometry teaches you to think with precision,
To analyze problems and make decisions.
It hones your mind, sharpens your wit,
And prepares you for challenges you'll meet.

From Pythagoras to Euclid's art,
Geometry has played a vital part.

Architects, artists, and engineers,
All use its principles to conquer their fears.
    So, dear student, embrace this subject true,
Let Geometry guide and inspire you.
For within its depths, you'll discover the key,
To unlock the wonders of the world and be free.

FIFTEEN

GUIDE YOUR QUEST

In the realm of shapes and lines,
Where logic and beauty intertwine,
Lies the subject of Geometry,
A gateway to a world of glee.
    With angles acute and angles obtuse,
Triangles, circles, and their use,
Geometry teaches us to perceive,
The hidden patterns we can achieve.
    Through the study of shapes so grand,
We learn to reason and understand,
To solve puzzles with method and grace,
Geometry unveils its magical embrace.
    From Euclid's theorems to Pythagoras' delight,
Geometry guides us with profound insight,

It sharpens our minds, expands our view,
Unlocking secrets that are hidden from few.

    So, dear student, do not dismay,
Geometry leads you on a remarkable way,
Embrace its challenges and dive right in,
For in its depths, a world of wonders begin.

    With perseverance and a curious mind,
You'll uncover treasures, rare and kind,
Geometry, like a compass, will guide your quest,
To pass this subject, be your very best.

    For in the realm of shapes and lines,
Lies the power to shape your own designs,
Geometry, a language to explore,
A journey that will forever endure.

SIXTEEN

GEOMETRY'S SECRETS

In the realm of shapes and lines,
Where logic so divinely shines,
There lies a subject, strong and true,
Geometry, waiting for you.

With angles acute and obtuse,
And circles, perfect and profuse,
Geometry beckons, don't you see?
A gateway to logic's mastery.

In proofs and theorems, you shall find,
A world of reason, one of a kind,
For Geometry's language, clear and pure,
Teaches you how to think and endure.

Unlock the doors to spatial grace,
Embrace the challenge, set the pace,

For in this world of angles and sides,
A student's spirit truly resides.

    So fear not the triangles and squares,
For Geometry's beauty, it shares,
With those who dare to take the leap,
To conquer the subject, oh so deep.

    Let your mind soar, like a bird in flight,
Through geometric wonders, day and night,
And as you pass this test with glee,
Geometry's secrets, you shall see.

    For in the realm of shapes and lines,
A world of knowledge, so divine,
A student's journey, bold and bright,
Geometry leads you to the light.

SEVENTEEN

GEOMETRY'S WISDOM

In the realm of shapes and lines,
Where logic dwells and beauty shines,
Lies the world of Geometry,
A subject that holds the key.

Oh, student, hear my humble plea,
Embrace this art, set your mind free.
For Geometry teaches not just math,
But the grace and power in every path.

Through angles, triangles, and curves,
You'll learn to think, your mind observes,
The patterns hidden in the world around,
In nature's symphony, a harmonious sound.

From Pythagoras to Euclid's hand,
Geometry's wonders forever expand,
Unlocking secrets, shaping your own designs,
A world of possibilities, where brilliance combines.

So study hard, with passion and zest,
For Geometry's challenges put to the test,
But fear not, dear student, for within its core,
Lies the magic of shapes, forevermore.

Let compass guide you, ruler in hand,
As you navigate this mystical land,
For in Geometry's embrace you'll find,
A universe of knowledge, one of a kind.

So rise, young student, reach for the sky,
Let Geometry's wisdom help you fly,
And as you pass this subject with pride,
A world of logic and beauty will be your guide.

EIGHTEEN

KNOWLEDGE INTERTWINE

In the realm of lines and shapes so fine,
Where logic meets grace, and wonders entwine,
Geometry, oh student, holds the key,
To unlock the world's hidden mysteries.

With angles acute and angles obtuse,
The mind is sharpened, brought to use,
For geometry teaches us to see,
The patterns woven in reality.

From triangles to circles, polygons grand,
Each figure holds a lesson at hand,
Symmetry, congruence, and proportion's art,
Geometry's language, a testament of smart.

For in this subject, you shall find,
A gateway to knowledge of every kind,

The power to reason, to analyze,
To see the world through discerning eyes.
  So let not frustration cloud your view,
Embrace the challenge, let curiosity brew,
For geometry's beauty lies in its might,
To shape your mind and guide you right.
  Pass this subject, and you shall see,
The wonders of geometry, forever free,
A world of logic, grace, and design,
Where the boundaries of knowledge intertwine.

NINETEEN

YOU'LL FIND THE KEY

In the realm where angles dance and lines entwine,
Lies a subject that can truly redefine.
Geometry, the art of shapes and space,
A path to logic, reason, and grace.

Beyond the numbers and equations it holds,
A world of patterns and secrets unfolds.
With compass and ruler, we explore,
The vast landscapes of Euclidean lore.

Triangles, circles, and polygons galore,
Unveiling the wonders we can't ignore.
From Pythagoras' theorem to the golden ratio,
Geometry reveals nature's grand tableau.

Angles teach us to measure and compare,
To analyze and reason, to think with care.

Proofs and theorems, like puzzles to solve,
Sharpening our minds as they evolve.

So fear not, dear student, embrace the quest,
Geometry's challenges will put you to the test.
For within its realm, you'll find the key,
To unlock the wonders of math and set your mind free.

With perseverance and a curious eye,
You'll soar through the subject, reaching the sky.
Geometry's power, it will ignite,
And guide you to success, shining so bright.

So take heart, dear student, don't be dismayed,
The beauty of Geometry will never fade.
With patience and effort, you'll surely surpass,
And conquer the subject with brilliance and class.

TWENTY

POWER TO SHAPE

In the realm of shapes and lines, behold,
Where Geometry's wonders unfold.
A subject that challenges the mind,
With logic and reasoning intertwined.

Geometry, the language of the divine,
Where patterns and symmetry align.
Through angles and curves, we find grace,
In every triangle, circle, and space.

A student's journey, a quest to explore,
With compass and ruler, they'll soar.
Unlocking the secrets, uncovering the key,
To a world of knowledge, vast and free.

For Geometry, dear student, holds the key,
To unlock the wonders that lie within thee.

With every theorem and proof you learn,
Your mind expands, your horizons turn.
    Embrace the challenge, don't be dismayed,
For in Geometry, beauty is displayed.
From Pythagorean theorem to the golden ratio,
A world of wonders, waiting to bestow.
    So study hard, with passion and might,
And conquer Geometry's lofty height.
For in its embrace, you'll surely find,
The power to shape your own design.

TWENTY-ONE

TRIANGLES, CIRCLES, AND POLYGONS UNITE

In the realm of shapes and lines, where angles intertwine,
There lies the magic of Geometry, a subject so divine.
Beyond the theorems and formulas that grace its mighty name,
Lies a world of logic and reasoning, where brilliance finds its flame.

Geometry, oh wondrous art, teaches us to see,
The symmetry and patterns that shape our reality.
With compass and ruler, we navigate the unknown,
Unlocking the secrets that lie within, like a treasure yet unshown.

For in the realm of Geometry, grace does surely dwell,
As shapes dance and twirl, creating a symphony so swell.
Triangles, circles, and polygons unite,
In a harmonious blend of elegance and light.

So, dear student, fear not the challenges that await,
For Geometry's embrace will lead you to a higher state.
Let your mind soar, like a bird in the sky,
And discover the wonders of Geometry, as time goes by.

With perseverance and determination, you shall succeed,
And Geometry's mysteries, you will indeed exceed.
So, take heart, dear student, for within you lies the key,
To unlock the power and beauty of Geometry.

TWENTY-TWO

ILLUMINATE YOUR WAY

In the realm of shapes, where angles reside,
Geometry waits, a subject to guide.
With compass and ruler, we journey along,
Unveiling the mysteries, capturing the song.
    Oh, student of numbers and curves so precise,
Unlock the secrets, let your mind suffice.
For Geometry teaches logic and reason,
A language of patterns, a mathematical season.
    Through triangles and circles, squares and lines,
A dance of symmetry, where grace intertwines.
For every figure, there's a story to tell,
A puzzle to solve, a knowledge to compel.
    Embrace the challenge, let your spirit soar,
For Geometry's power, you can't ignore.

It shapes the world, from buildings to art,
A foundation of knowledge, a vital part.

    So study with passion, with eyes open wide,
Discover the wonders that Geometry hides.
Let it guide your footsteps, illuminate your way,
For in this subject, lies the path to success, they say.

    Oh, student of Geometry, let your mind unfurl,
And conquer the subject, like a radiant pearl.
For in this journey, you'll find your true might,
And pass the test, shining ever so bright.

TWENTY-THREE

PURE DELIGHT

In the realm of shapes, where lines entwine,
Geometry's wisdom, so divine.
A subject of logic, reason, and grace,
Unlocking mysteries, in every embrace.

Oh, student, listen, don't be dismayed,
For Geometry's beauty will never fade.
It's not just triangles, squares, and spheres,
But a language that whispers in your ears.

In angles and arcs, a dance takes flight,
A symphony of numbers, pure and bright.
With compass and ruler, you'll find your way,
Through the labyrinth of shapes, day by day.

For Geometry teaches you to observe,
To analyze, deduce, and gently curve.

It hones your mind, sharpens your sight,
Guiding you through the darkest night.

    So, study hard, don't let it blur,
Geometry's secrets, you'll soon uncover.
Embrace the challenge, don't be afraid,
For with perseverance, success will be made.

    Let curiosity be your guiding light,
In this geometric journey, shining so bright.
Pass the subject, let your dreams take flight,
For Geometry's embrace, is pure delight.

TWENTY-FOUR

FOREVER EXPLORE

In the realm of lines and curves, where shapes take form,
Lies the power to unlock secrets, and embrace the norm.
Geometry, a language, spoken by the wise,
A gateway to logic, reasoning, and the skies.
    With compass in hand, and ruler in tow,
Discover the wonders that Geometry bestows.
Angles and triangles, circles and spheres,
Unveil the beauty that the eye reveres.
    For in this realm, grace finds its place,
Symmetry and balance, a harmonious embrace.
In the dance of polygons, elegance does bloom,
Geometry, a muse, in every classroom.
    Embrace the challenge, let your mind take flight,

Through proofs and theorems, seek the light.
For Geometry's lessons, they go beyond measure,
They shape your mind, and bring you pleasure.

    So fear not the numbers, the angles, the lines,
For Geometry's wisdom, it forever shines.
Unlock the doors, let your spirit soar,
Pass this subject, and forever explore.

TWENTY-FIVE

DEAR STUDENT

In the realm of lines and angles,
Where logic unfolds its wings,
Geometry, the queen of shapes,
Unveils the secrets she brings.

Oh, student, hear my humble plea,
Embrace this subject with grace,
For in the world of geometry,
Your mind shall find its rightful place.

Through circles, triangles, and squares,
The language of shapes takes flight,
Unlocking doors to realms unknown,
Guiding your thoughts like a guiding light.

For geometry is more than numbers,
It teaches you to reason and think,

To solve puzzles with precision,
And dance on the edge of the brink.

    So fear not the challenge it presents,
Let curiosity be your guide,
For within the realm of geometry,
Boundless wonders reside.

    Discover symmetry's perfect embrace,
And the elegance of each design,
See beauty in the shapes you trace,
Let your imagination intertwine.

    With a compass and a ruler in hand,
You hold the power to create,
Geometry, the key to dreams,
Unleash your potential, do not wait.

    So, dear student, let your spirit soar,
Embrace the subject with all your might,
For in the realm of geometry,
A world of infinite possibilities takes flight.

TWENTY-SIX

SEIZE THE DAY

In the realm of shapes and lines,
Where logic intertwines,
Lies the beauty of Geometry,
A subject that holds the key.

With compass and ruler in hand,
A student's journey will expand,
Unlocking the secrets of space,
And finding grace in every trace.

Geometry, the language of form,
Teaches us to think and transform,
It sharpens our mind's keen sight,
Guiding us through the darkest night.

From circles to triangles, squares to spheres,
Geometry wipes away all our fears,

It builds a bridge from old to new,
And clears the path for dreams to pursue.

    So, student, let your spirit soar,
Embrace Geometry, forever more,
For within its realm, you will find,
A world of wonders to unwind.

    With perseverance, you shall prevail,
Geometry's lessons will never fail,
For in its depths, you'll come to see,
The beauty of shapes and the power to be.

    Pass this subject, seize the day,
Geometry will light your way,
So, spread your wings and take the flight,
Into a world where knowledge ignites.

TWENTY-SEVEN

GEOMETRY INSPIRE

In the realm of shapes and lines,
Where logic and beauty intertwine,
Lies the power of Geometry,
Unlocking secrets for all to see.
    A student's journey, filled with grace,
In this subject they must embrace.
For Geometry teaches, with patient might,
The art of reasoning, shining bright.
    Through angles and curves, they'll learn to perceive,
The world's hidden patterns, they shall achieve.
Triangles, circles, and polygons rare,
Hold the key to knowledge beyond compare.
    With compass and ruler as their guide,
They'll navigate the geometric tide.

They'll measure, calculate, and explore,
Unveiling wonders never seen before.

So, dear student, fear not the unknown,
For Geometry is a garden full-blown.
Pass this subject, and you shall prevail,
With a mind sharpened, ready to set sail.

Let the allure of Geometry inspire,
To reach new heights, to aim much higher.
With perseverance and dedication true,
Success in this subject shall be your due.

So, study hard, with passion and zest,
Geometry's challenges, you'll pass the test.
For within its realm, you'll surely find,
A world of wonders, to expand your mind.

TWENTY-EIGHT

SOUL COLLIDE

In the realm of shapes and lines,
Where logic meets the divine,
Lies the beauty of Geometry,
Unlocking secrets for all to see.
    Oh, student, do not despair,
For Geometry holds wonders rare.
In its realm, you'll find the key,
To unlock the door of harmony.
    Geometry teaches you to think,
To reason and to never shrink.
It builds a bridge from mind to hand,
Creating structures that will stand.
    With triangles, squares, and circles bright,
You'll learn to see with clearer sight.

Angles, proofs, and theorems galore,
Will open doors you've not explored.
   Embrace the challenge, don't shy away,
In Geometry's realm, you will sway.
Discover the magic within its grasp,
And watch as your knowledge starts to clasp.
   So, dear student, study with zeal,
For Geometry's wonders are real.
Pass this subject with grace and pride,
And let your mind and soul collide.

TWENTY-NINE

LESSONS ARE PROFOUND

In the realm of shapes and lines,
Where logic and beauty intertwine,
Lies the subject of Geometry,
A gateway to knowledge, full of mystery.

Embrace this world of angles and curves,
For within its depths, wisdom preserves.
Unlock the secrets it holds within,
And watch your understanding begin.

Geometry teaches us to reason,
To see the patterns in every season.
From triangles to circles and polygons,
We learn to solve problems and carry on.

With compass and ruler in hand,
There's an elegance in every command.

For in Geometry, we find grace,
In the symmetries that interlace.
    So, dear student, do not despair,
Embrace the challenge, show you care.
For Geometry holds a treasure trove,
Of knowledge, logic, and beauty to behove.
    Pass this subject with determination,
And open the doors to imagination.
For Geometry's lessons are profound,
In shaping minds and futures unbound.

THIRTY

LABYRINTH OF SHAPES

In the realm of shapes and lines,
A world of wonders you shall find,
Geometry, a subject divine,
With secrets waiting to unwind.

Unlock the doors of logic's gate,
Where reason and order contemplate,
Angles, triangles, circles, too,
They'll guide your mind, help you break through.

For in this realm, grace is born,
In every curve and angle, it's sworn,
Symmetry and balance intertwine,
Creating beauty, so sublime.

So fear not the geometric way,
Embrace the challenge, seize the day,

With compass, ruler, and a steady hand,
You'll conquer Geometry's demanding land.
   For in this labyrinth of shapes,
A treasure trove of knowledge awaits,
The power to reason, to create,
To see the world in a different state.
   So, student, let your passion ignite,
Geometry will be your guiding light,
For in its depths, you'll surely see,
The brilliance of your own geometry.

THIRTY-ONE

STUDY WITH PASSION

In the realm of shapes, where lines entwine,
Lies a subject called Geometry, divine.
Oh, student, listen, let me tell you true,
The wonders it holds, the dreams it can imbue.

Geometry, the language of form and space,
Unlocks the secrets of this wondrous place.
It teaches logic, reason, and grace,
A foundation on which success we embrace.

From Euclid's compass to Pythagoras' theorem,
Geometry weaves a tapestry, a radiant gleam.
Triangles, circles, polygons so vast,
Their beauty and elegance forever will last.

With angles acute and angles obtuse,
The mind expands, creativity is set loose.

For in this subject, you'll find a key,
To shape your thoughts, your destiny.
    So fear not the challenge, embrace it tight,
For in conquering Geometry, you'll find your light.
Unlock the doors to a world unknown,
Where logic and beauty together are sown.
    So study with passion, with fervor and glee,
Geometry's treasures await, for you to see.
Pass this subject, and you'll find your way,
To a future bright, where dreams hold sway.

THIRTY-TWO

DEDICATION

In the realm of shapes and lines,
Where logic and beauty intertwine,
Lies the enchanting world of Geometry,
A subject that holds the key.
    Oh, student, fear not the challenge ahead,
For Geometry's mysteries can be easily spread.
It teaches you to reason and to think,
To solve puzzles that make your mind blink.
    With angles and triangles, you'll learn to see,
The symmetries that exist both far and near.
And when you grasp the power of shapes,
A world of possibilities it creates.
    Geometry, the language of form,
Guides us through life's intricate storm.

It teaches us grace and harmony,
Unveiling truths for all to see.
  So, brave student, open your mind,
Embrace the beauty you will find.
For Geometry is not just a subject to pass,
It's a gateway to a world vast.
  Unlock the secrets it holds within,
And let your journey with Geometry begin.
With dedication and a curious heart,
Success in this subject shall be your art.

THIRTY-THREE

BRIGHTEN YOUR DAY

In the realm of shapes and lines, where logic reigns supreme,
Lies the gateway to a world where brilliance gleams.
Geometry, the language of the universe's design,
Unfolds the mysteries that lie within its confines.
    With angles acute and obtuse, triangles and squares,
Geometry teaches us to reason and compare.
For every problem we solve, a puzzle we unravel,
We sharpen our minds, our intellects, and travel.
    Through proofs and theorems, we learn to think clear,
To analyze, deduce, and persevere.
For in Geometry, we find the grace so divine,
The elegance of patterns, like poetry in each line.
    Beyond the classroom walls, its applications grow,

From architecture's marvels to nature's stunning show.
For Geometry is the foundation, the key,
Unlocking secrets, shaping our own destiny.

So, dear student, embrace this subject's might,
For in its challenges, lies infinite delight.
Let the beauty of shapes guide your way,
And Geometry's triumphs will brighten your day.

Pass this test of angles and curves with pride,
For in Geometry, endless wonders reside.
With logic, reasoning, and grace as your guide,
Success in this subject, you shall surely stride.

THIRTY-FOUR

NEVER PASS YOU BY

In the realm of shapes and lines,
Where angles dance and curves align,
Lies a subject, both art and science,
That beckons you to take a chance.

Geometry, a puzzle to unravel,
A key to unlock the mind's travel,
For in its depths, you'll find the treasure,
Of logic, reasoning, and boundless measure.

Like a painter with brush in hand,
Geometry allows you to command,
The canvas of space, with grace and precision,
Creating symphonies of mathematical vision.

From Pythagoras to Euclid's lore,
Geometry's secrets we adore,

For it unveils the patterns of the universe,
And teaches us to seek, explore, and converse.

So fret not, dear student, in this endeavor,
Geometry's challenges, you shall conquer,
Embrace the lines and angles with delight,
And watch as your understanding takes flight.

For in the realm of shapes and space,
Geometry holds a special place,
A language of patterns, both old and new,
That will guide you to vistas you never knew.

So embrace the challenge, hold your head high,
Geometry's wonders will never pass you by,
And as you pass this subject's test,
A world of possibilities will manifest.

THIRTY-FIVE

GEOMETRY'S REWARDS

In the realm of shapes, where angles meet,
Lies a subject that's both simple and sweet.
Geometry, the language of lines and curves,
Teaches us lessons that our mind preserves.

Unlock the doors to logic's grand domain,
Where reasoning and proof forever reign.
For in Geometry's sacred realm we find,
The power to expand our thoughtful mind.

With compass and ruler, we measure and draw,
Creating patterns that leave us in awe.
Triangles and circles, squares and more,
Geometry's beauty we truly adore.

In every structure that catches our gaze,
Geometry's hand guides its elegant ways.

From towering skyscrapers to bridges so grand,
Geometry's principles form the firmest stand.

So, dear student, fear not this noble art,
Embrace its challenges, let them impart,
The grace and precision that dwell within,
And watch as your understanding begins.

For in the realm of Geometry's embrace,
Lies a world of wonders, a magical space.
So, study hard, with passion and zeal,
And let Geometry's secrets gently reveal.

Pass this subject, let your spirit soar,
For Geometry's knowledge is worth much more.
Open your mind, let curiosity thrive,
And Geometry's rewards will surely arrive.

ABOUT THE AUTHOR

Walter the Educator is one of the pseudonyms for Walter Anderson. Formally educated in Chemistry, Business, and Education, he is an educator, an author, a diverse entrepreneur, and he is the son of a disabled war veteran. "Walter the Educator" shares his time between educating and creating. He holds interests and owns several creative projects that entertain, enlighten, enhance, and educate, hoping to inspire and motivate you.

Follow, find new works, and stay up to date
with Walter the Educator™
at WaltertheEducator.com

www.ingramcontent.com/pod-product-compliance
Lightning Source LLC
LaVergne TN
LVHW052000060526
838201LV00059B/3748